Forces and Motion

Contents

Preview the Book 2
What Are Forces? 3
   Pushes and Pulls 4
   Friction 6
   Gravity 7
   Magnetic Force 9

How to Read Charts 10
How Do We Describe Motion? 11
   Position and Motion 12
   Speed 14
   Velocity 16
   Acceleration 17

Cause and Effect 18
What Are Newton's Laws of Motion? 19
   The Science of Motion 20
   Newton's First Law of Motion 21
   Newton's Second Law of Motion 22
   Newton's Third Law of Motion 23

Glossary 24

Preview the Book

You read nonfiction books like this one to learn about new ideas. Be sure to look through, or *preview*, the book before you start to read.

First, look at the title, front cover, and table of contents. What do you guess you will read about? Think about what you already know about forces and motion.

Next, look through the book page by page. Read the headings and the words in bold type. Look at the pictures and captions. Notice that each new part of the book starts with a big photograph. What other special features do you find in the book?

Headings, captions, and other features of nonfiction books are like road signs. They can help you find your way through new information. Now you are ready to read!

What Are Forces?

MAKE A CONNECTION

Skydivers jump from an airplane and fall toward Earth. Do you know what makes things fall? What makes a skydiver slow down when his or her parachute opens?

FIND OUT ABOUT

- how forces can change the motion of objects
- what happens when many forces act on one object at the same time
- friction, gravity, and magnetic force

VOCABULARY

force, p. 4

friction, p. 6

gravity, p. 7

weight, p. 8

magnetic force, p. 9

◀ A soccer player can make a ball go faster and farther by applying greater force.

Pushes and Pulls

A **force** is needed to make an object move. You apply a force when you push, pull, throw, or lift an object. A force can

- make an object start moving
- make a moving object speed up, slow down, or stop
- change the direction in which an object moves

When you apply more force to an object, the object's motion changes more. A gentle kick will make a ball move slowly and go a short distance. A hard kick will make that ball move quickly. It will go much farther.

You need to use more force to move objects with more mass. Mass is the amount of material in an object. It takes more force to start a full shopping cart moving than an empty one. That is because the full cart has more mass.

A full shopping cart has more mass than an empty one. So it takes more force to get a full cart moving. ▶

▲ The team on the left is pulling harder. So the forces on the rope are unbalanced. The rope moves in the direction of the greater force.

Many different forces can act on one object at the same time. Some forces may be stronger than others. Some may act in different directions. *Net force* is the combination of all the different forces acting on an object. Think of two teams playing tug-of-war. The way the rope moves depends on the combination of forces acting on it.

When both teams are pulling the same amount, the rope does not move. That is because the forces acting on it are *balanced*. When forces are balanced, the net force is zero.

When one team pulls harder than the other, the rope starts moving. That is because the forces acting on it are *unbalanced*. So the net force is not zero. The force in one direction is greater than the force in the other direction. The rope moves in the direction of the greater force.

✓ Name two objects that you used force to move today. Which object took more force? Why?

▲ Adding oil to machine parts can lower friction.

▲ Some bike tires have rough tread patterns that make friction greater. The tires do not slip as much on a trail.

Friction

A force called **friction** happens when two objects touch, roll, rub, or slide against each other. Friction acts against an object's motion. Friction can stop an object from moving, or it can slow down a moving object.

The amount of friction depends on the surfaces that are touching. Rough surfaces often make more friction than smooth surfaces. That is why gravel is less slippery to walk on than smooth ice. The gravel makes more friction with the bottom of your shoes.

Sometimes we want less friction. It can cause machine parts to slow down and wear out. We often can lower friction by adding oil or grease.

Sometimes friction can be helpful. It lets us grip things more tightly. We can make friction greater by using rough or sticky materials.

 How does friction change the way things move?

Gravity

Gravity is a force that acts between all objects that have mass. The force of gravity pulls objects toward each other. Gravity even acts across great distances. For example, gravity keeps the planets in orbit around the Sun. Gravity keeps the Moon in orbit around Earth.

Gravity is a very important force. Gravity keeps us on the ground. It makes things fall to Earth unless something holds them up.

Gravity is stronger between objects that have more mass. Earth has a huge amount of mass. So the force of gravity between a person and Earth is easy to notice. The force of gravity between a person and a desk is much weaker. But it is still there!

Gravity also is stronger between objects that are closer together. When there is more distance between objects, the gravity between them is less.

◀ A juggler applies a force to toss a ball. Gravity slows the ball, changes its direction, and makes it fall back down.

Weight is a measurement of the force of gravity on an object. Maybe you have used a scale to find an object's weight in pounds (lb). In science, we describe forces in units called newtons (N). Weight is the force of gravity. So weight can be described in newtons. The weight of a small apple is about 1 N (about 0.22 lb).

Weight and mass are not the same. Mass is the amount of material in an object. We measure mass with a balance. Mass is often described in grams (g) or kilograms (kg).

Remember that the force of gravity between objects depends on their mass. So your weight on Earth depends on the mass of both you and Earth. You would weigh less on the Moon than on Earth. That is because the Moon has less mass than Earth. So the force of gravity pulling on you would be weaker on the Moon.

 Gravity is stronger between objects that are closer together. How do the masses of objects affect the force of gravity between them?

▲ A spring scale is one way to measure the weight of an object.

◀ Your Moon weight would be about one-sixth your Earth weight. Divide your Earth weight by six to find your Moon weight.

▲ A magnet exerts a force on some metal objects.

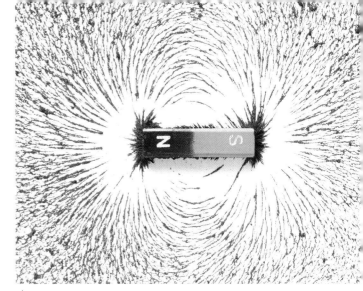

▲ Small bits of iron can help you see the magnetic field of this bar magnet.

Magnetic Force

Magnets can pull on, or attract, certain objects. The force of a magnet is called **magnetic force**. Objects made of the metals iron, nickel, or cobalt are attracted to a magnet. Magnets also attract most steel objects. This is because steel has iron in it. A magnet can move an object without even touching it.

The effect of a magnet on the area around it is called its magnetic field. The magnetic field gets weaker with distance from the magnet. The magnetic field is strongest at the places on the magnet called poles. A bar magnet has a magnetic pole at each end. The poles are often labeled *N* for north-seeking and *S* for south-seeking. Unlike, or opposite, poles attract each other. Like poles push each other away, or repel. So an *N* and an *S* pole will attract. But two *N* poles will repel.

 What is one reason that a paper clip might *not* stick to a magnet?

REFLECT ON READING

You previewed pictures, captions, and other book features before reading. Which book features were most helpful when you read about forces? How did they help you?

APPLY SCIENCE CONCEPTS

Think about throwing a basketball. It goes up in the air, through the basket, and falls to the floor. Tell what forces act on the ball as it moves.

How to Read Charts

A chart is a way to summarize or compare information. Charts group facts into columns and rows. Columns go down the chart. Rows go across the chart. Charts are sometimes called tables.

Nonfiction books like this one often contain charts. You will find a chart on page 15.

TIPS

Follow these steps when you read a chart.

1. Read the title.
2. Read the headings across the top and down the left side of the chart. Be sure you understand what they mean.
3. To look for information, read both across *and* down. Use your finger or a ruler to find the place where the correct row and correct column meet.
4. Read the caption.

How Do We Describe Motion?

MAKE A CONNECTION

Racecars zoom around and around a track. What could you say about the position and motion of the cars?

FIND OUT ABOUT

• ways we tell about the position and motion of objects

VOCABULARY

position, p. 12

motion, p. 12

distance, p. 13

speed, p. 14

velocity, p. 16

acceleration, p. 17

Position and Motion

You can talk about your location, or **position**, by comparing it to other objects. If someone asks where you are, you might say that you are in a classroom. Or, you might say that you are next to a table.

Motion happens when an object changes position. Forces cause motion. Push a toy car, and it goes in a straight line. Spin a top, and it turns around many times. How do you tell if an object is moving? Follow and measure its position over time.

Think about riding a bicycle. How do you know that you are in motion? You change position compared to houses, trees, and other things that are standing still. Those things are your *frame of reference*.

▲ Suppose you were standing by the lake in this picture. How could you tell if the bike riders were in motion?

◀ The girl can tell the bus is moving. Her frame of reference includes still objects outside the bus.

Your frame of reference has a lot to do with how you see motion. Think about sitting on a bus. If you look at only the inside of the bus, you may think the bus is not moving. But if you look out the window, your frame of reference changes. You see that you are changing position compared to buildings. So you know that the bus is moving.

You can measure the distance something moves. **Distance** is how far it is from one point to another. We often measure distance in units such as inches, feet, yards, and miles. In the metric system, distance is measured in units called millimeters (mm), centimeters (cm), meters (m), and kilometers (km).

The units we use depend on the distance we are measuring. Millimeters and centimeters are used for shorter distances. Meters and kilometers are used for longer distances.

✔ How can you tell if an object is in motion?

▲ You can measure short distances with a ruler or a tape measure.

13

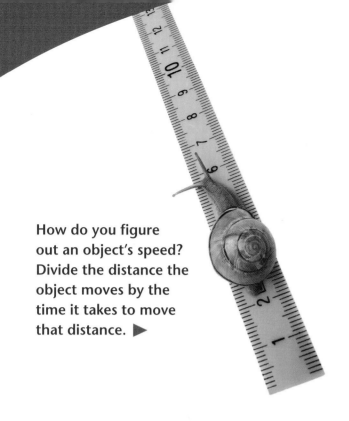

How do you figure out an object's speed? Divide the distance the object moves by the time it takes to move that distance. ▶

▲ You can use a stopwatch to measure the time it takes a runner to move a distance.

Speed

Speed is a measure of how quickly an object's position changes. You can tell how quickly an object moves by measuring its speed. Some objects move so slowly that it is hard to tell if they are moving at all. Other objects move very quickly.

To find an object's speed, you need to know two things.

1. the distance the object moves
2. the amount of time it took to move that distance

Then, you divide the distance by the time:

$$\textit{speed} = \textit{distance} \div \textit{time}$$

Say you walked a distance of 10 kilometers (about 6 miles) in a time of 2 hours. You could find your speed this way.

$$\textbf{10 km} \div \textbf{2 hr} = \textbf{5 km/hr}$$

This means that you walked at a speed of 5 kilometers (about 3 miles) per hour.

Objects do not usually move at the same speed over a long distance. So it is helpful to talk about their average speed for an entire journey.

Think about a flock of birds flying north in the spring. The birds stop and start many times during their trip. Their speed changes from day to day. Suppose the birds fly about 6,437 km (about 4,000 mi) in about 31 days. Their average speed during the journey is about 208 km (about 129 mi) per day.

 You need to know two measurements in order to find an object's speed. What are they?

Average Speeds

Animal or Object	Speed (metric)	Speed (customary)
Snail	0.045 km/hr	0.028 mi/hr
Sloth	0.13 km/hr	0.08 mi/hr
Galapagos Tortoise	0.3 km/hr	0.19 mi/hr
Sphinx Moth	53 km/hr	33 mi/hr
Cheetah, running	100 km/hr	62 mi/hr
Fastest Baseball Pitch	168.7 km/hr	104.8 mi/hr
Peregrine Falcon, diving	320 km/hr	198 mi/hr
Jet, average cruising speed	920 km/hr	572 mi/hr
International Space Station	28,000 km/hr	17,000 mi/hr

▲ Which animal or object has the fastest average speed? The slowest?

Velocity

Velocity is a measure of both speed *and* direction. Say you are riding in a car that is going 50 kilometers per hour (about 31 miles per hour). You are going north. Your velocity is 50 kilometers per hour, north.

Two objects moving at the same speed can have different velocities. Say your friend's car is also going 50 kilometers per hour. But your friend is going south. Your velocity is 50 kilometers per hour, north. Your friend's velocity is 50 kilometers per hour, south.

Velocity lets you figure out the position of moving objects at any time. Think about flying a spacecraft to another planet. You would need to know the speed *and* direction of both the moving planet and your spacecraft.

 How can two things move at the same speed but have different velocities?

The velocity of these runners is 20 kilometers per hour, east. ▶

A roller coaster car often changes speed and direction. So it is almost always accelerating.

Acceleration

Acceleration is any change in an object's velocity. An object accelerates if it slows down, stops, speeds up, or changes direction. You may have heard people use the word *accelerate* to mean only "speed up." In fact, objects also accelerate when they slow down or turn.

An object accelerates when a net force acts on it. When you kick a ball, it accelerates in the direction of the kick. You can apply a greater force by kicking harder. Then the ball will accelerate more.

It takes more force to accelerate objects with more mass. Think again about pushing a shopping cart. You could make an empty cart roll with a small push. But you would have to push a full cart harder. That is because it has more mass.

 Name three ways something can accelerate.

REFLECT ON READING

Look again at the chart on page 15. Suppose you added a row to the chart for your favorite animal's average speed. Where might the row go? How can you find out your animal's actual speed?

APPLY SCIENCE CONCEPTS

Write in your science notebook about how your position and motion both change on your way to school. Be sure to tell about your speed, velocity, and acceleration.

Cause and Effect

A **cause** is the reason something happens. An **effect** is what happens as a result of the cause.

On page 21 of this section, you will read about a property of objects called inertia. Think about the effects of inertia on the motion of objects.

TIPS

Thinking about causes and effects can help you understand why things happen.

- To find effects, ask, "What happens?"
- To find causes, ask, "Why does this happen?"
- Look for signal words such as *cause*, *effect*, *because*, *why*, *therefore*, *since*, *so*, and *as a result*.
- A cause may have more than one effect. An effect may have more than one cause.

A cause and effect chart can help you keep track of your ideas about why things happen.

cause	→	effect

What Are Newton's Laws of Motion?

MAKE A CONNECTION

Rowers pull their oars through the water. Why do you think the boat moves one way as the oars move the other way?

FIND OUT ABOUT

- the scientific study of motion
- three important principles that help us explain and predict the motion of objects

VOCABULARY

inertia, p. 21

The Science of Motion

People have tried to understand the motion of objects for thousands of years. In ancient Greece, philosophers thought about motion. They came up with ideas about why things move.

In time, scientists added to the Greeks' ideas by doing experiments. Galileo Galilei (1564–1642) was a scientist from Italy. He rolled balls down ramps and tossed objects into the air. He tied weights to a rope to see how they would swing. Galileo wrote about his experiments and ideas about motion.

Sir Isaac Newton (1642–1727) was a scientist from England. He read about Galileo's work. He watched how things in nature move. He also worked with ramps and other machines. Newton explained that the motion of objects follows three basic laws. The laws sum up many ideas you have just read about.

 How did Galileo and Newton study moving objects?

Experiments with ramps led to some important ideas about motion. ▼

▲ Sir Isaac Newton wrote about three laws of motion. The laws help us predict what happens when forces act on objects.

▲ Bowling pins have inertia. They will not move until a net force acts on them.

▲ Suppose a car stops suddenly. The riders keep moving forward because of inertia. Seat belts apply the force needed to stop the riders and help keep them safe.

Newton's First Law of Motion

An object at rest will not move until a net force acts on it. An object that is moving will keep moving at the same speed and in the same direction until a net force acts on it.

Newton's first law of motion tells us that no acceleration happens without a net force. It takes a net force to make something start moving, speed up, slow down, change direction, or stop.

Newton's first law is also called the law of inertia. **Inertia** is a property of objects. It tells how objects keep moving or stay still until a net force acts on them. A still object has inertia. The object will not move until a net force acts on it. So, for example, bowling pins will not move until a ball hits them.

Moving objects also have inertia. A moving object will keep moving at the same speed and in the same direction until a net force acts on it. So a rolling ball will keep moving until a force stops it. The force might come from another object. Or it might come from friction alone.

 Tell how a book sitting on your desk has inertia.

Newton's Second Law of Motion

An object's acceleration depends on the amount of force acting on the object and on the object's mass. An object acted on by a net force will accelerate in the direction of the net force.

Newton's second law of motion tells us how acceleration, force, and mass relate to one another. When more force acts on an object, the object accelerates more. Think about hitting a tennis ball. If you give the ball a light tap, it accelerates slowly. If you hit the ball very hard, it accelerates very fast.

Objects with more mass need more force to accelerate. Think about a full shopping cart rolling along level ground. It would take more force to stop the full cart than to stop an empty cart. That is because the full cart has more mass.

An object will accelerate in the direction of the net force acting on it. Remember the tug-of-war from page 5. One team pulled harder than the other. The rope moved in the direction of the greater force.

 Tell how an object's acceleration depends on its mass and the force acting on it.

◀ The more force a tennis player uses, the more the ball accelerates.

▲ When a skateboarder's foot pushes on the ground, the ground pushes back on the skateboarder's foot.

▲ When an oar pushes the water, the water pushes back on the oar.

Newton's Third Law of Motion

> *For every action force acting on an object, the object will apply an equal and opposite reaction force.*

Newton's third law tells us that forces always come in pairs. When one object applies a force to a second object, the second object applies a force right back. The first object's force is the action force. The second object's force is the reaction force. The reaction force is always equal to the action force in strength. It is always opposite in direction.

Think of pushing yourself along on a skateboard. One foot pushes against the ground. That's the action force. The ground pushes against your foot. That's the reaction force. The reaction force acts on *you*. It is strong enough to push you forward. Your action force acts on *Earth*. The action and reaction forces are the same strength. But the strength of your force is too small to make Earth move. This is because Earth has a very large mass.

 What are action and reaction forces?

REFLECT ON READING

Make a cause and effect chart like the one on page 18. Write "inertia" in the cause box. What effect does inertia have on a moving object? Add this to the chart.

APPLY SCIENCE CONCEPTS

Jump up and down in place. Use Newton's laws of motion to tell about how forces act on your body when you jump.

Glossary

acceleration (ak-sel-uh-RAY-shuhn) any change in an object's velocity **(17)**

distance (DIS-tens) how far it is from one place to another **(13)**

force (FORS) a push or pull that acts on an object **(4)**

friction (FRIK-shuhn) a force that happens between objects that are touching and acts against, or opposes, their motion **(6)**

gravity (GRAV-uh-tee) the force that acts between objects that have mass, causing them to pull toward, or attract, each other **(7)**

inertia (in-UR-shuh) a property that tells how objects tend to keep moving or stay still until acted on by a net force **(21)**

magnetic force (mag-NET-tik FORS) the force of a magnet **(9)**

motion (MOH-shuhn) a change in the position of an object **(12)**

position (puh-ZISH-uhn) the location of an object **(12)**

speed (SPEED) the measure of an object's change in position over time **(14)**

velocity (vuh-LAHS-uh-tee) the measure of an object's speed in a certain direction **(16)**

weight (WAYT) the strength of the force of gravity on an object **(8)**